I0469040

Seeing the True Sun

by Rolf A. F. Witzsche

Text Copyright (c) 2015 - Rolf A. F. Witzsche

All rights reserved

Contents

About the Illustrated Science series
*On the Ice Age and Climate Change
and the book*

Seeing the True Sun

What do we see in the mind when we look at the Sun? That's a basic question. Do we see a blazing fireball powered by nuclear fusion within, in a super-hot environment where hydrogen gas is fused into helium? If this is how the Sun is seen, the viewer is deceived by a fairy tale. Fairy tales are often chosen as bed-time stories to put children to sleep. How about seeing the true Sun? It has the potential to wake society up.

With the next Ice Age on the near horizon, potentially beginning in the 2050s, which cannot be recognized in the land of fairy tales, the recognition of the true nature of the Sun, that drives the Ice Age dynamics, becomes an existentially critical issue. Being 'asleep,' we may die in the easy chair when the glaciation conditions resume.

Plasma in the physical universe is as challenging in perception as the spiritual domain in the human sphere. Both are invisible, except by their effects, but they are understandable and knowable. But how does one break away from the fairy tales that inspire delusions? Answers must be found.

With the Ice Age Challenge now before us, we face two imperatives. One is to understand the real physical dynamics that power and affect the Sun, and with it to create the physical infrastructures that enable human living to continue in an Ice Age climate. The second challenge, and this is the greater challenge, is to raise up our humanity to such height as will impel us to get the job done. Some say that miracles are needed on both fronts. But what of it? Are we, as human beings, not the miracle makers on the Earth?

In the real universe, the cosmic operations are anti-entropic in nature, and expanding and progressing. We, ourselves are evidence of this progression. Should this progression have ended? Neither is our Sun isolated from the progressive nature of the universe, but expresses its dynamics, its resonating plasma streams, and their reflection in the climate on Earth. Shouldn't we develop ourselves spiritually and culturally, likewise?

Climate Change reflects the nature of the universe. It should also be reflected in us.

The Earth itself is the creation of the Sun, with its atoms having been massively synthesized in high-energy times near the center of the galaxy.

The synthesizing plasma fusion is presently at a low state, though it is currently enhanced for our Sun by electromagnetic 'Primer Fields' that focus interstellar plasma onto the Sun in a highly condensed manner. When the plasma-focusing system becomes inactive, below the required threshold conditions, the Sun reverts to a type of cosmic default level with 70% less energy being radiated, and higher rates of solar cosmic-ray flux being experienced.

At the present rate of plasma diminishment being experienced, the solar activity phase-shift threshold to the next Ice Age period may be crossed in 30 years, or in the 2050s, most likely. With the primer-fields system gone inactive by then, the climate on Earth will get 40 times colder than the Little Ice Age in the 1600s had been. Ice core evidence promises that. Without the needed preparations for human living in such an environment, 99% of humanity would die of starvation, both by the cold, and by CO_2 depletion that diminishes agriculture, as more CO_2 becomes dissolved into the sea.

With the 'Primer Fields' being critical for our very existence, the exploration of them is likewise critical.

In the Little Ice Age, between 10% and up to 30% of the populations in Europe had perished by starvation. The last Big Ice Age was evidently vastly harsher. Only 1-10 million people emerged from it alive. That's all we had after 2 million years of development. We want to do far better this time around; and we can, with large-scale technological infrastructures for our food supply. But will we create them? Will we get the job done in the 30 years that we still have left before the Ice Age starts anew? Will we even consider it? And how certain are we that the phase shift to the next glaciation period will begin, as the evidence

6

suggests, in the 2050s? We have no slack on this front. Should we fail us on this absolute front, we would be committing suicide.

Numerous fields of evidence tell us that the next Ice Age is near. That's where the truth begins. Most of the evidence was discovered in the 1990s and thereafter. Some evidence is measured in ice cores; some is measured in space, by satellites. Some measurements are also made on the ground in terms of measurements of the Earth's magnetic-pole drift observed in northern Canada. All of this is seen combined with high-energy physics experiments at a leading national laboratory, and is also explored in the small in static experiments.

So, what will the answer be? Will we move with the evidence? Or will we lay ourselves down to die by default?

It takes an independent researcher to brake the taboos that have kept mainstream cosmology imprisoned, increasingly, during the past century, even while what is regarded as taboo is known to be wrong.

The Illustrated Science series is intended to open the scene beyond the threshold of accepted taboos, to where the actual physical evidence speaks for itself.

The scope of the existential challenge that the Ice Age brings with it, takes astrophysics out of the academic domain and places it into the foreground as one of the most-critical issues of our time. The big Climate Change events that have already worldwide effects are mere fringe effects in the flow of the ever-changing cosmic dynamics. The big effect, when the Ice Age begins anew, promises to be caused by a dimmer and colder Sun. The loss of 70% of the Sun's radiated energy defines our climate future that begins in the near term.

Sure, we can live with all that by creating new platforms for agriculture that are able to operate under Ice Age conditions. But will we do it? The task is enormous. Or will we fail ourselves on this front? We have no reason to allow us to fail. We have the materials and energy resources on hand to accomplish everything that is required for us to continue to live in an Ice Age World. But will we do it? The big question that never goes away, therefore, is; will we develop our inner resources as human beings sufficiently to get the job done, and to get it done in time? Or will we do nothing, ignore the challenge, and condemn our children and one-another to an agonizing death by starvation? That's the choice.

Towards meeting the inner challenge, I have created the epic series of novels, The Lodging for the Rose. And further, towards meeting the science challenge, I have produced numerous research books and several dozen exploration videos that the Illustrated Science series is modeled after. The work is the result of a quarter century of research, for which numerous elements of evidence in related fields came to light during the timeframe of my research.

It is my hope that the work that went into all of these projects will help in some degree - for humanity that we are all a part of - to write itself a ticket to have a future.

High-resolution color images, of the images in this book, can be obtained at www.iceagetheatre.ca

*How do we see the Sun when it illumines our world?

How do we see the Sun when it stands high and illumines our world? Do we see the truth? Or do we see its face as the fog-shrouded composite image of numerous fairy tales.

Grand fairy tales about the Sun are taught

Grand fairy tales about the Sun are taught in nearly all the schools. Students are even graded on their personal efficiency in learning the fairy tales. Those tales are many. Those who learn them well, earn themselves an A on their report card. That's the school song.

But do the students know anything

Corel corp.

But do the students in the end, including the A-students, know anything that is intrinsically true about the Sun, or resembles the truth even remotely?
Of course they don't. And why should anyone be worried about that?

Discovering the truth about our Sun

Discovering the truth, especially the truth about our Sun, hasn't been on the agenda in the schools for a long time. Thus, I need to repeat the question: How do we behold the Sun when it illumines our world? Do we see it how it really is? Hardly! We behold it in terms of the fairy tales that are taught.

*Let's review the fairy tales about the Sun: a sphere of hydrogen gas?

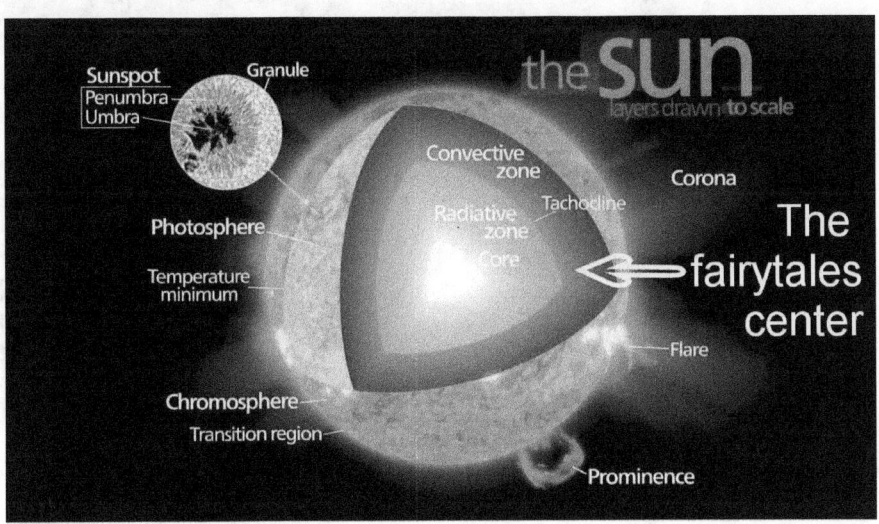

Let's review the fairy tales.

For starters, the Sun is deemed to be a large sphere of hydrogen gas that is more than a million times larger in volume than the earth, but contains only 300,000 times more mass. However, the enormous mass of it is deemed to be so large, acting on the core, that the gas becomes compressed in the core to a density that is 150 times greater than the density of water, which adds up to, 13 times the density of the element, led.

Something doesn't add up here.

With the Sun's core 13 times denser than led

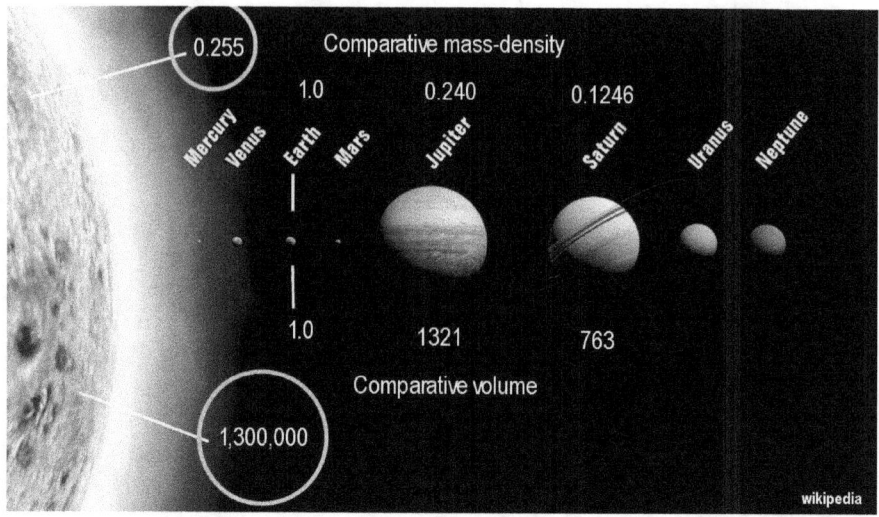

With the Sun having a core that is 13 times denser than led, the Sun should be 1000 times 'heavier' than it is known to be. Instead of it being immensely heavy, the measured mass-density of the Sun is roughly the same as that of Jupiter. With the core of the Sun being 13 times 'heavier' than led, the measured extremely light 'weight' of the Sun shouldn't be possible.

So, were can we find the truth? The truth is fairly obvious, isn't it? It is simply not possible for the Sun to be a sphere of hydrogen gas. The measured evidence precludes this option. If one looks at the evidence with the 'eye' of the mind, only one option remains for a superlight Sun to exist. This option is, that the Sun exists as a sphere of plasma.

Plasma is a 'soup' that contains the primary particles

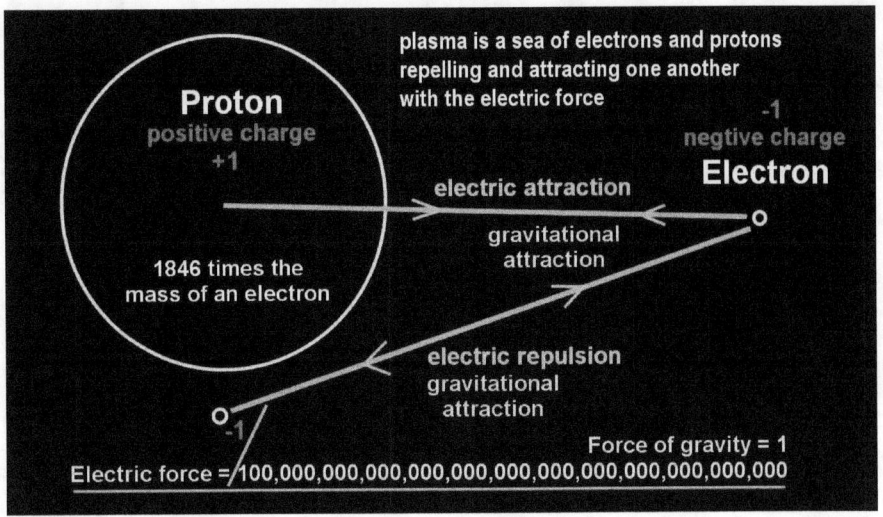

Plasma is a 'soup' that contains the primary particles that all atoms are made of. There exist two types of such particles, called protons and electrons. They are both electrically charged. Protons repel each other by the electric force, by their positive polarity, and so do electrons by the negative polarity. However, both attract one another. The ratio between them, between their mutual repulsion and attraction, determines to a large degree a plasma's density.

The balance of the forces in plasma

Super Giant Star/Sun

dwarf Giant Star

the Sun among stars

Comparison of stars - CC BY 3.0 wikipedia - ESO/M. Kornmesser

Left to right: a red dwarf, the Sun, a blue dwarf, and R136a1

The balance of the forces in plasma enables a super-light Sun to exist, even a super-giant sun, which is not possible otherwise.

The evidence pins the 'label of truth' onto the plasma-type Sun.

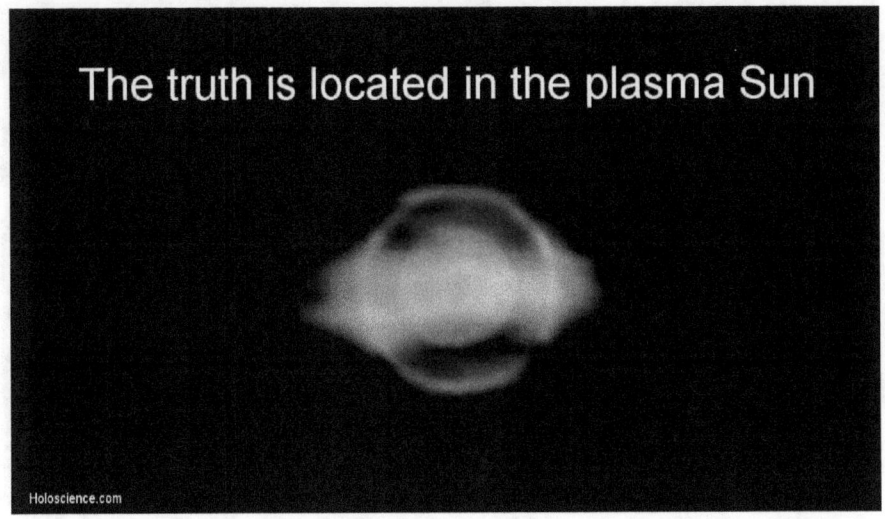

The evidence that this is so, pins the 'label of truth' onto the plasma-type Sun.

Let's get back to the fairy tales

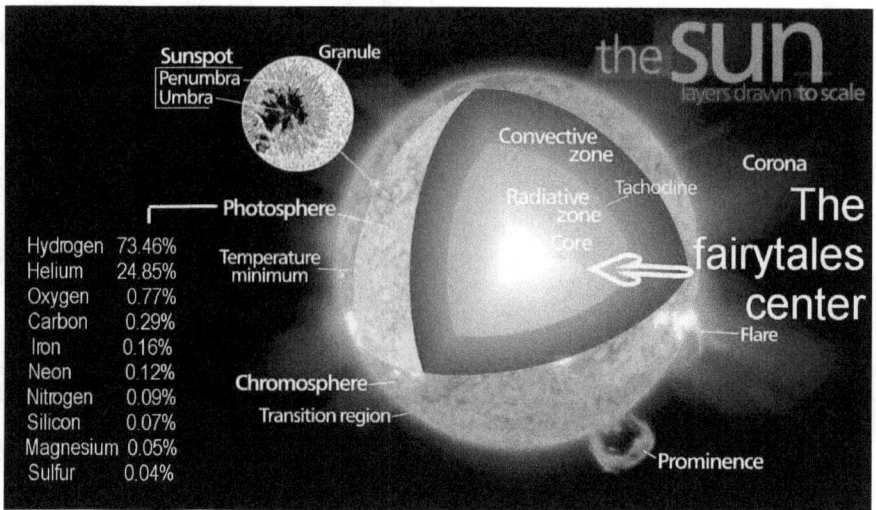

Let's get back to the fairy tales. Many more fairy tales have been written about the Sun.

One of the great fairy tales is that the Sun is heated internally, by nuclear fusion occurring in its core that combines hydrogen atoms into helium atoms, and generates energy in the process.

Since the Sun cannot be a sphere of hydrogen gas, this tale is not really possible, is it? We have ample evidence that the impossible doesn't actually happen in the real Sun.

Sunspots erupt that are holes ripped into the solar surface

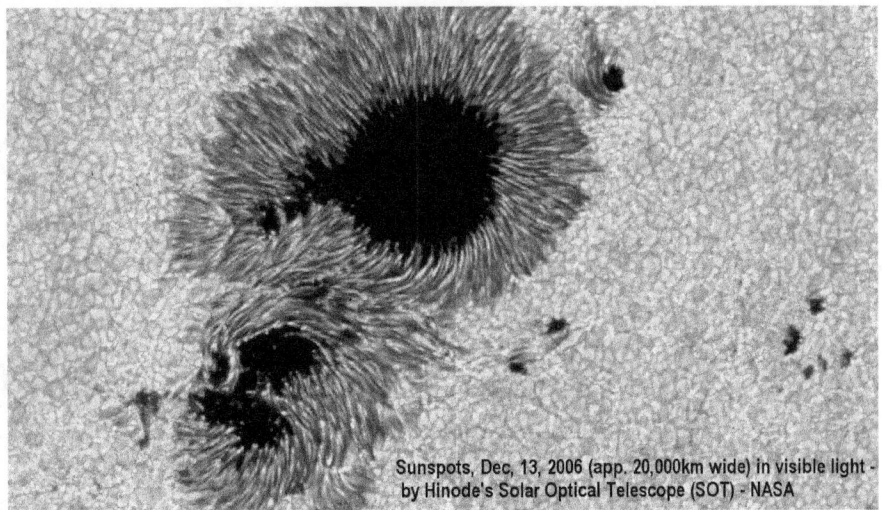

Sunspots, Dec, 13, 2006 (app. 20,000km wide) in visible light - by Hinode's Solar Optical Telescope (SOT) - NASA

That no giant reactor lurks under the surface of the Sun, deep within it, becomes evident when sunspots erupt that are holes ripped into the solar surface.

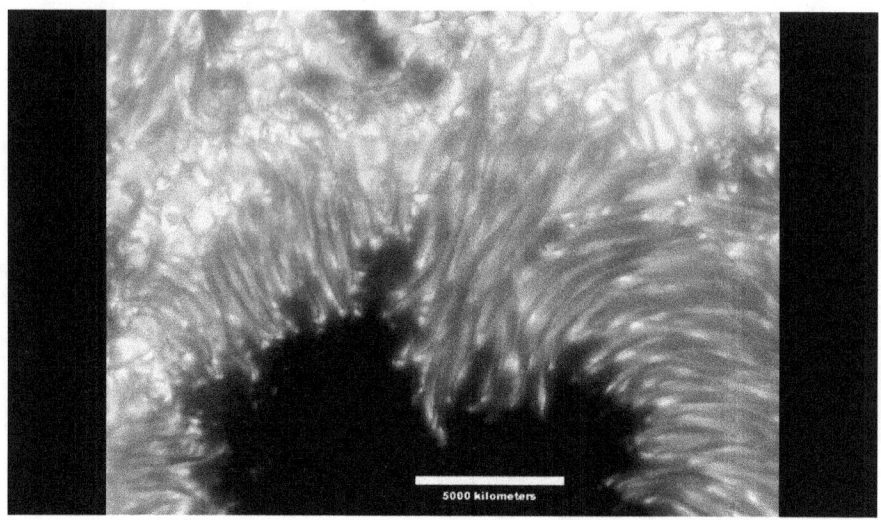

5000 kilometers

By looking through the umbra of the sunspots, we see a dark sun below the surface in every case. The Sun reveals itself thereby as being essentially dark and empty inside.

Solar surface a plasma-fusion reactor

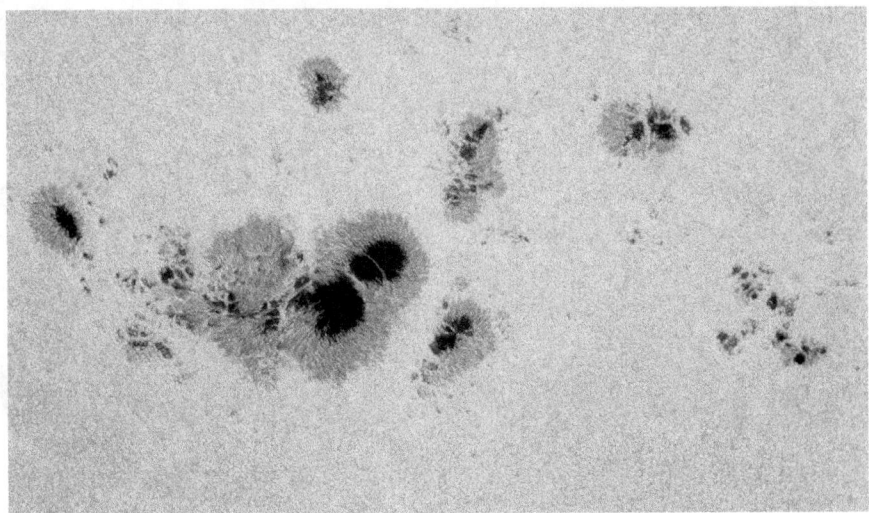

This doesn't mean that nuclear fusion is not occurring on the Sun. It only means that the fusion reactions occur on the surface of the Sun, not within it. On the surface, interstellar plasma is fused into atomic elements by electric concentration and acceleration of interstellar plasma onto the Sun. The entire solar surface thereby becomes a plasma-fusion reactor.

The resulting, rather plainly apparent evidence

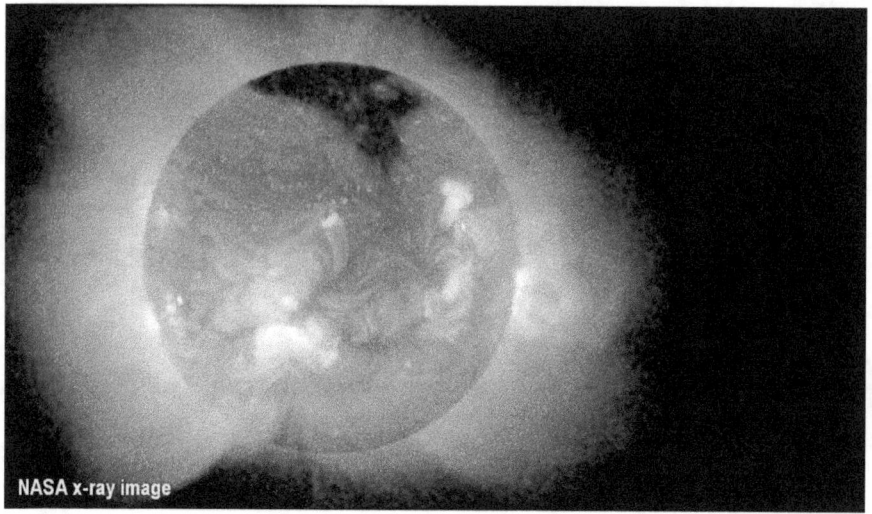

NASA x-ray image

The resulting, rather plainly apparent evidence, pins another 'label of truth' onto the plasma-type Sun.

The fairy tale has still another deep flaw

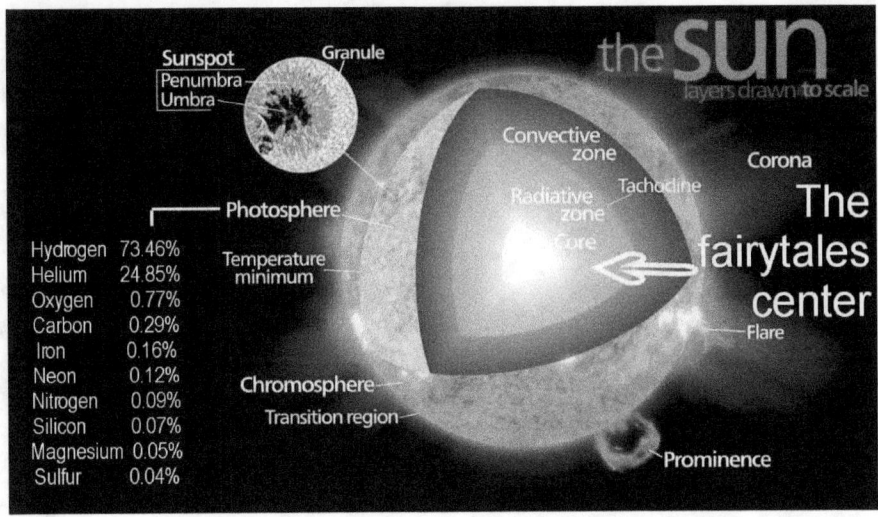

Now let's get back to the fairy tale, for more of its imaginary stories that are far removed from the truth.

The fairy tale about the Sun has still another type of deep flaw. The flaw is, that in a tightly confined nuclear fusion cell, the fusion product dilutes the fusion fuel and snuffs the process out.

Every fusion reactor on Earth has this problem

Every fusion reactor on Earth has this problem. The longest fusion burn achieved to-date, by the machine shown here, lasted just slightly under a single second. And that's the world record, the best result that has been accomplished

The Sun wouldn't burn long on this basis

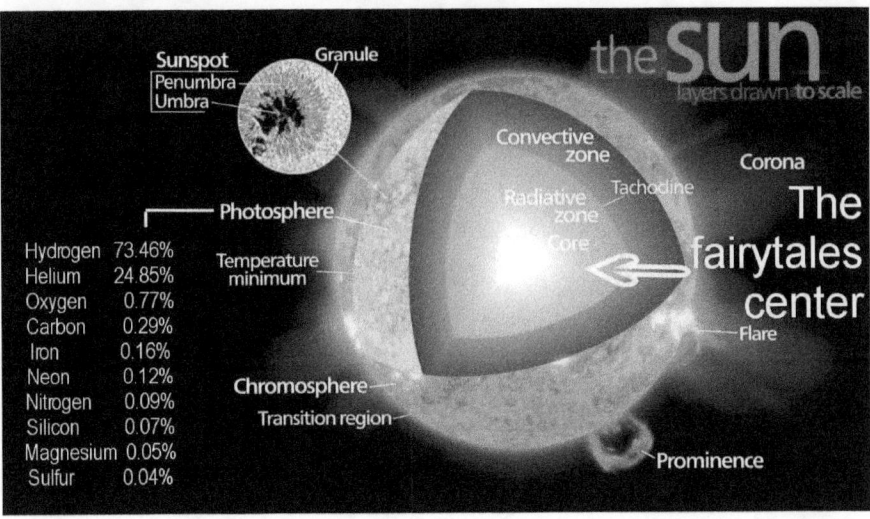

Obviously, the Sun wouldn't burn long on this basis, with its fusion product clogging up its reactor at the core. Nor is it possible for the generated helium to become vented away from the core. The heavy helium would have to work its way to the surface against the force of the massive solar gravity, across half a million kilometers of solar mass. The obvious fact is that the purging of the core cannot be happening.

Now consider the equally obvious fact

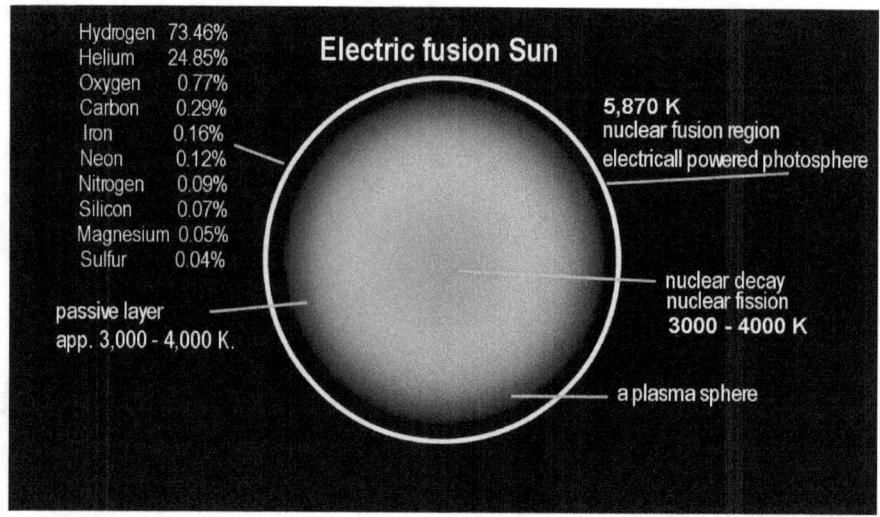

Now consider the equally obvious fact that the purging of the fusion product is not a big issue for the plasma-type Sun where the nuclear fusion reactions occur on the surface.

Obvious facts pin another' label of truth' onto the plasma Sun.

The truth is located in the plasma Sun

NASA x-ray image

These obvious facts, of course, pin another' label of truth' onto the plasma Sun.

So what about the solar fairy tale then?

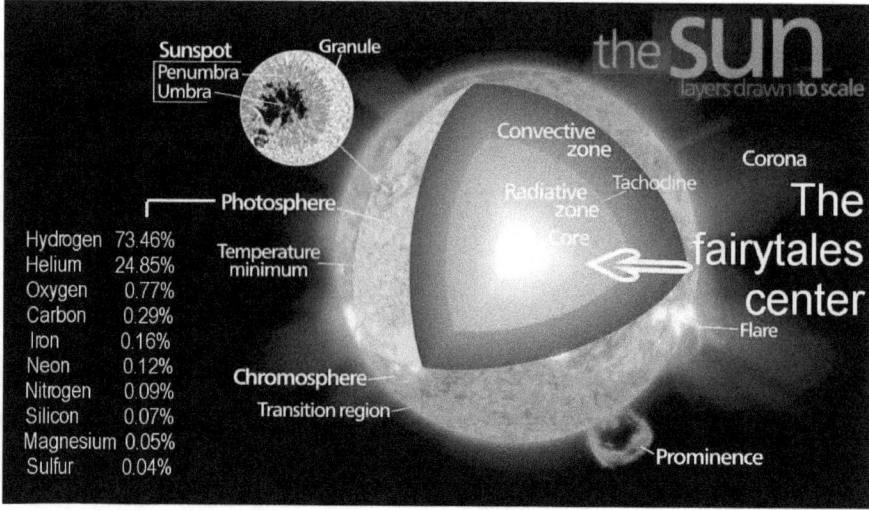

So what about the solar fairy tale then? Is this all there is? Oh, there are more stories imbedded, stories of sheer fantasy.

*The Sun as a hydrogen sphere is a paradox

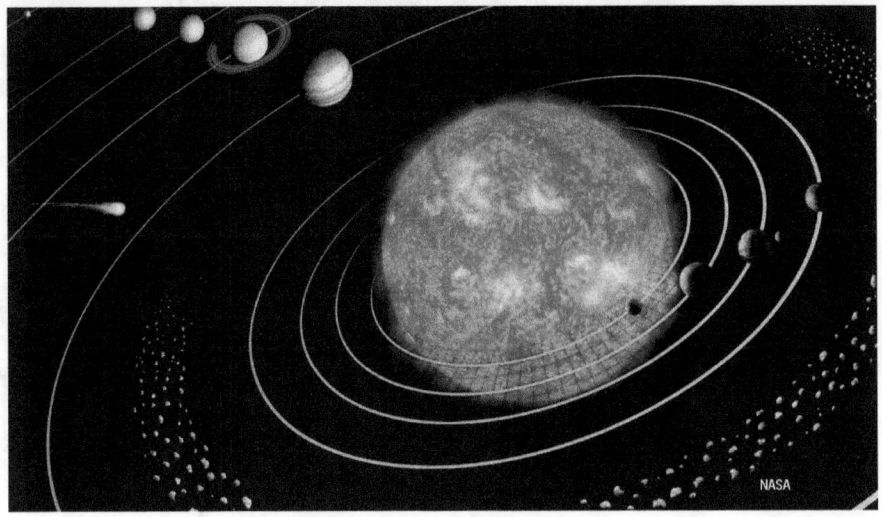

One of these stories is the hydrogen paradox. The Sun as a hydrogen sphere is a paradox. We are told that the Sun is the product of gravitational condensation of a cloud of primordial atomic elements that originated in the Big Bang explosion 13.6 billion years ago. If this was so, by what miracle, would the hydrogen atoms separate themselves out from the rest, and congregate to become a sun?
Shouldn't the Sun, by its enormous gravity have captured all the heavy elements according to the condensation theory, instead of all the heavy elements being found contained in the planets? Gravitational condensation doesn't happen the way the fairy tales tell the story. The evidence is totally different.

The impossible paradox is not a paradox in the plasma universe

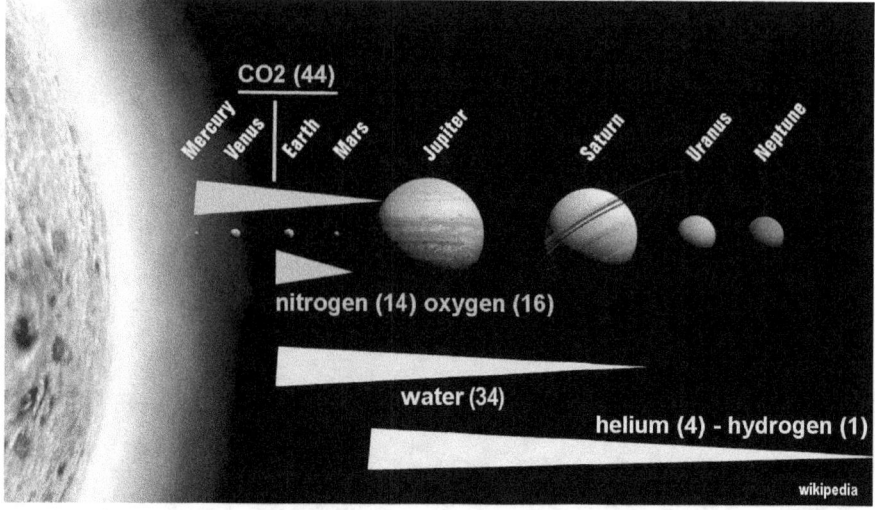

Of course the impossible paradox that the makeup of the solar system presents to us, is not a paradox in the plasma universe where a sun is forged from a high-density concentration of plasma that enables nuclear fusion reactions to occur on its surface where all the atomic elements in the solar system are synthesized, and have been synthesized for as long as the Sun existed. The synthesized elements are carried with the solar wind. The heavy elements, of course, fall out first and form the inner planets. This pattern is also evident with the heavy gases, in principle, being predominant across the inner planets

All the atomic elements actively created by a sun

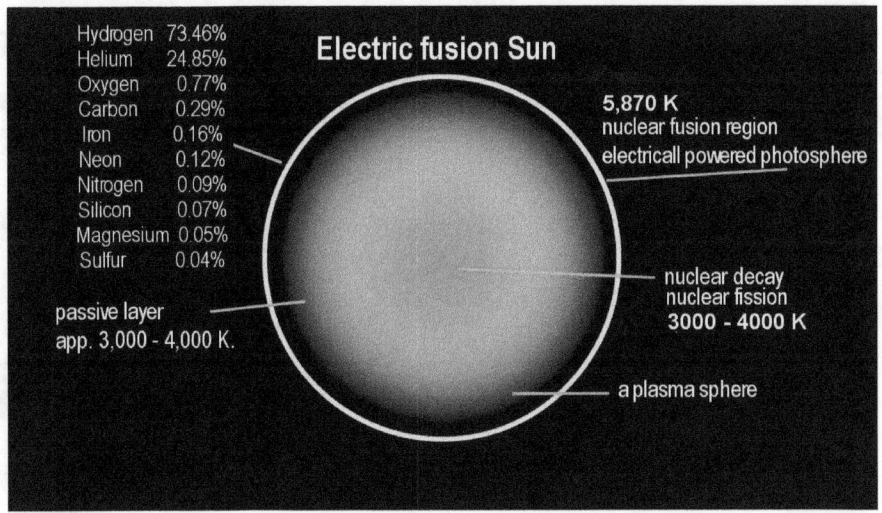

This is the reason why we see all the major atomic elements being present in the Sun's atmosphere. We see them there, because they are being created there, from where they flow away with the solar wind.

All the atomic elements in the universe were actively created by a sun. Gravitational condensation of Big Bang fairy tale dust is not needed for the universe to exist, or is even possible.

The solar system pins the 'label of truth' onto the plasma Sun

Thus, the solar system as we have it, all by itself, pins the 'label of truth' onto the plasma Sun.

A number of other paradoxes

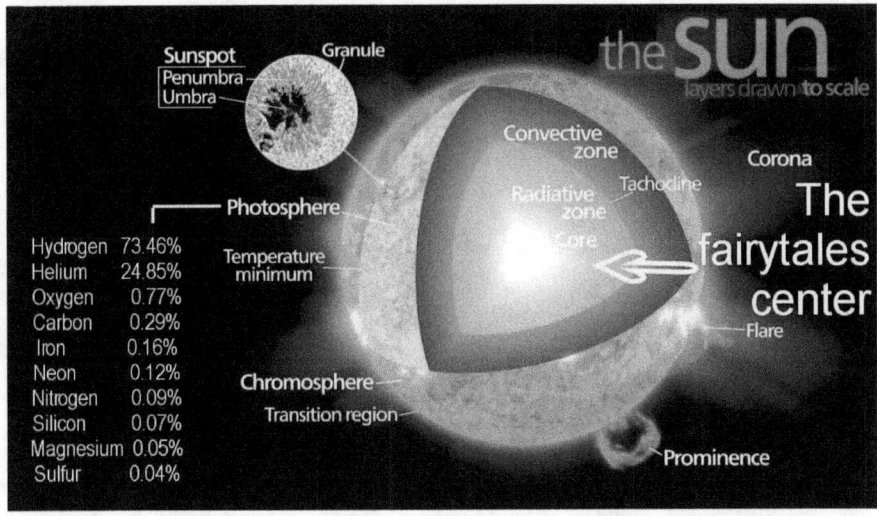

This brings us to a number of other paradoxes that the fairy tale doesn't even attempt to explain, as if it was some sort of miracle.

Solar wind accelerated away from the Sun, against the force of gravity

© Miloslav Druckmuller/Barcroft

http://www.zam.fme.vutbr.cz/~druck/Eclipse/ - an example of the amazing solar eclipse photography of Miloslav Druckmueller

It is a mystery in fairy tale land of what is called modern solar physics that the solar wind is being accelerated away from the Sun, against the force of gravity, to the enormous speed of up to 800 km per second. This shouldn't be happening if the Sun was internally powered.
Of course, if the Sun was powered by electric plasma interaction on its surface, the high-speed solar-wind acceleration is completely natural and is expected.

The process that accelerates the solar wind

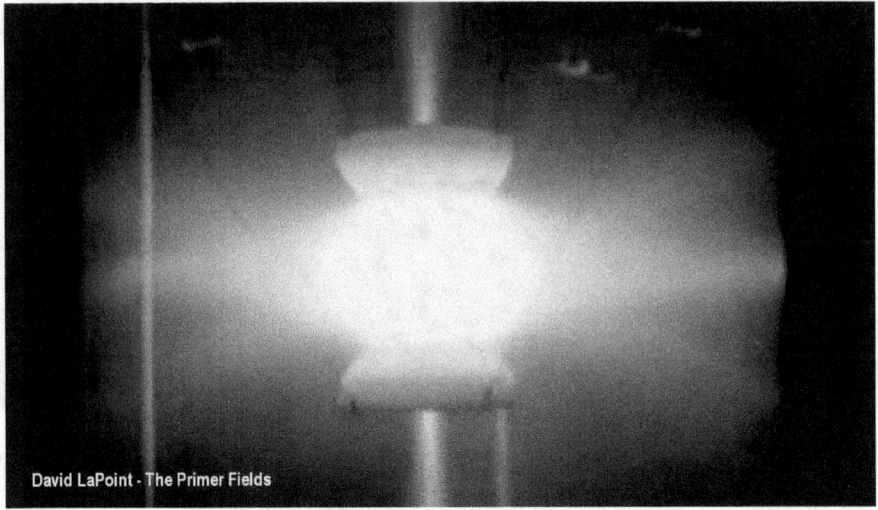

David LaPoint - The Primer Fields

The process that accelerates the solar wind is so natural that it can be replicated in principle in the laboratory, by replicating the function of the individual fusion cells that generate the solar wind.

The solar-wind paradox being resolved

The truth is located in the plasma Sun

NASA x-ray image

Thus, the solar-wind paradox being resolved, all by itself, pins the 'label of truth' once more onto the electrically powered plasma Sun. The plasma Sun doesn't have a paradox there. It solves the paradox.

The fairy tale includes still another paradox

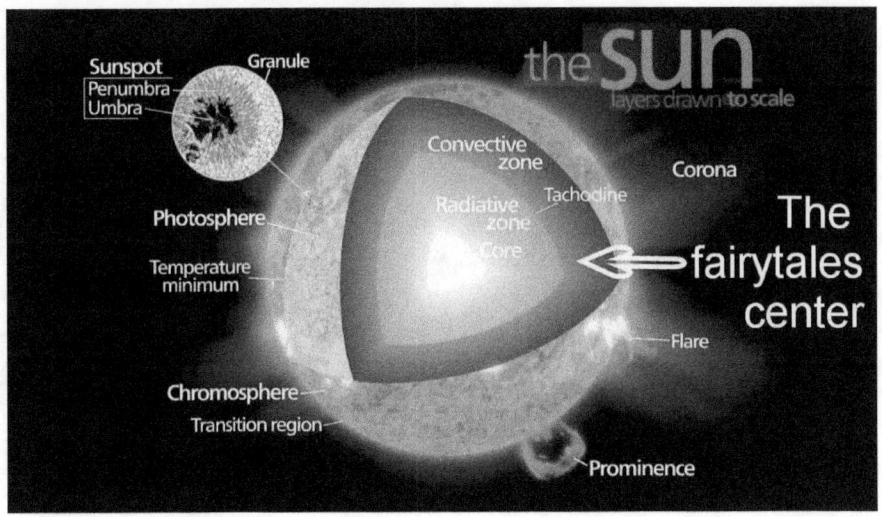

Now let's get back to the fairy-tale model once more, which includes still another paradox that is deemed a sort of miracle in the fairy tale.
The paradox is that of the Sun's super-heated corona.

The corona several hundred times hotter than the Sun

Solar corona   by Luc Viatour / www.Lucnix.be - wikipedia

The corona that surrounds the Sun, extending for long distances, is known to be several hundred times hotter than the surface of the Sun below it. The solar surface is measured at 5,500 decrees Celsius, while the corona has been measured in the range of millions of degrees. How is this possible if all the heat is deemed to come from the inside of the Sun? That's an impossible paradox to solve, or is it?

Of course, for a plasma sun the phenomenon of the super-heated corona is not paradoxical. Here, the solar corona is a mixture of atomic elements flowing away from the Sun, which are agitated by the electric interaction of flowing plasma streams that the atoms encounter, such as the high-speed solar wind, and the in-flowing plasma streams. What we see here simply means, that the heating of the solar corona is not a function of the heat emitted by the Sun, but is created by its own electric, dynamic process that is a typical aspect of a plasma Sun.

The corona too, pins the label of truth onto the plasma Sun

The truth is located in the plasma Sun

NASA x-ray image

Thus, the Sun's corona too, pins the label of truth onto the electrically powered plasma Sun. We always come back to that.

With so many labels of truth pinned onto the electric plasma Sun

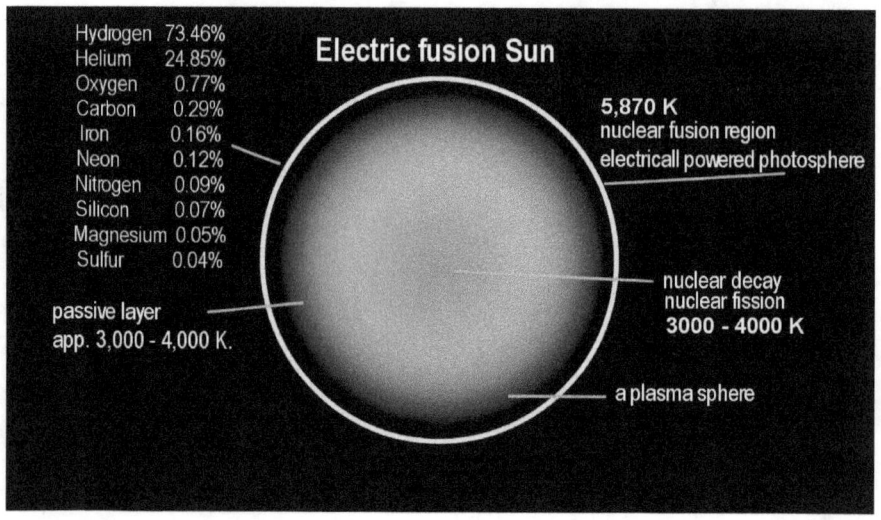

With so many labels of truth pinned onto the electric plasma Sun, it is self-evident that the truth lies outside the fairy-tale land of the internally powered Sun.

For the fairy-tale Sun no real evidence actually exists

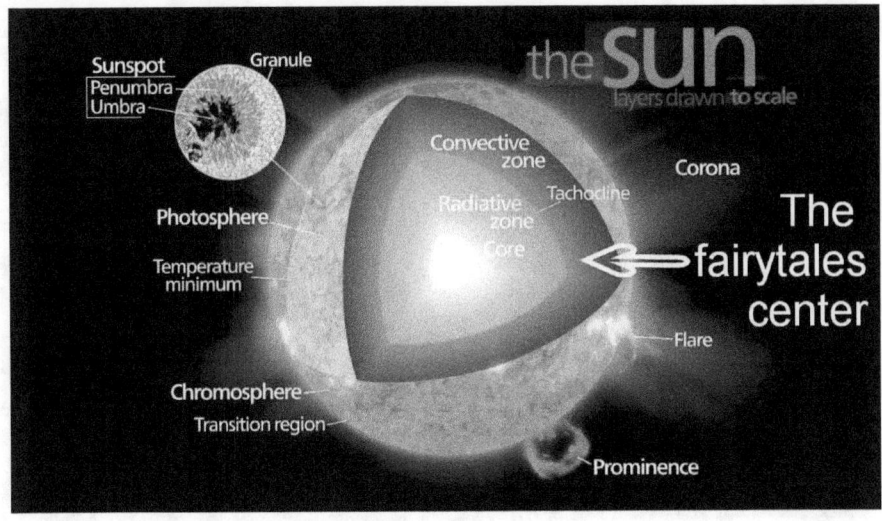

For the fairy-tale Sun no real evidence actually exists.

*Infinite contrast between the truth and the fairy tale

One may predict from this near infinite contrast between the truth and the fairy tale, that the truth will some day be taught in the schools and be acknowledged in society. The reason why this is presently not the case, is obviously political in nature.

The global warming doctrine

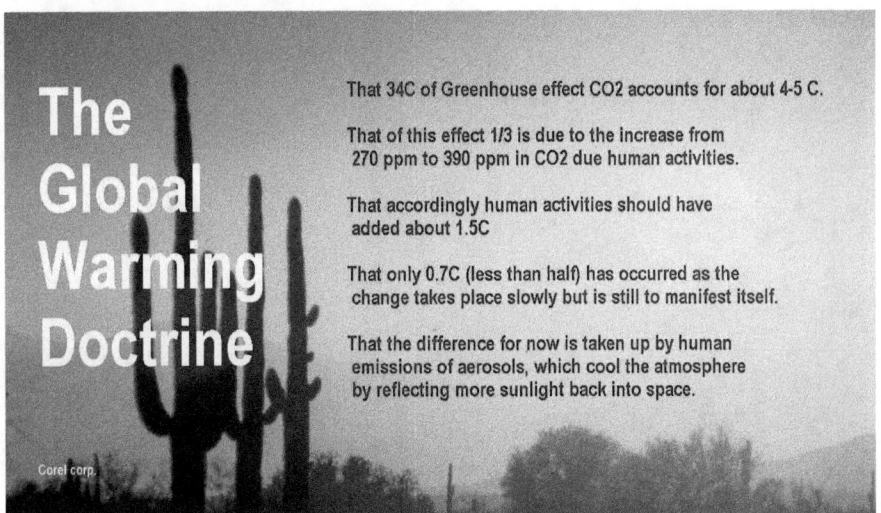

The
Global
Warming
Doctrine

That 34C of Greenhouse effect CO2 accounts for about 4-5 C.

That of this effect 1/3 is due to the increase from 270 ppm to 390 ppm in CO2 due human activities.

That accordingly human activities should have added about 1.5C

That only 0.7C (less than half) has occurred as the change takes place slowly but is still to manifest itself.

That the difference for now is taken up by human emissions of aerosols, which cool the atmosphere by reflecting more sunlight back into space.

Corel corp.

The global warming doctrine rests on the theory of the Sun being an invariable constant, for which the theory of the internally powered Sun is essential. With the Sun being deemed an invariable constant, it can be argued that all climate fluctuations on earth result from manmade causes, for which enormous penalties are now being inflicted on humanity, such as the destruction of industries and the mass-burning of food in a starving world.

The mass-burning of food in the form of biofuels

hE15 promotion Amsterdam   E85 in the USA

The mass-burning of food in the form of biofuels that are burnt in automobiles, is murdering up to 100 million people a year with starvation, as demanded by the depopulation policy in defence of the system of empire and its feudal platform. Thus, the entire green ideology and its devastating consequences is built squarely on a theory about our Sun that is demonstrably false in every respect.

The greatest danger of the false theory

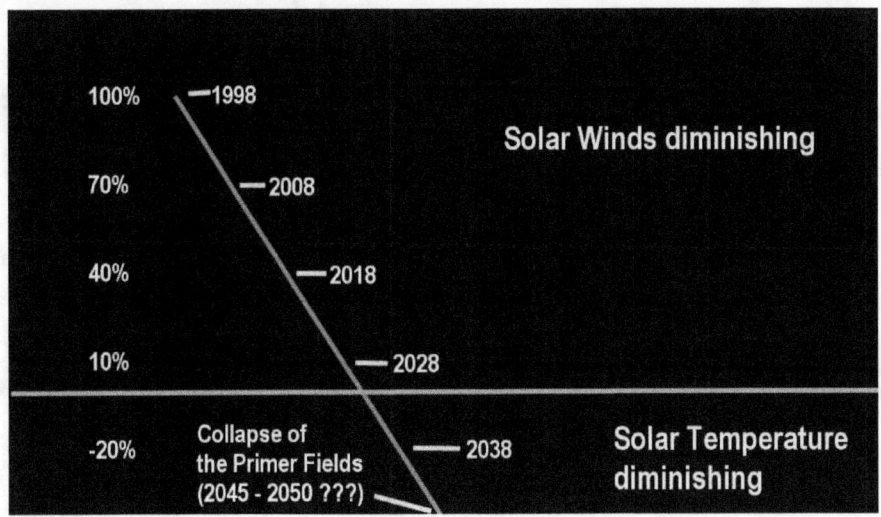

And this too, is only the fringe effect of the false theory. The greatest danger of the false theory is that it blocks the recognition of the presently unfolding solar dynamics in which we can see the start of the next Ice Age occurring in potentially the 2050s timeframe, with the Sun going inactive at this time.

The truth about the Sun remains blocked

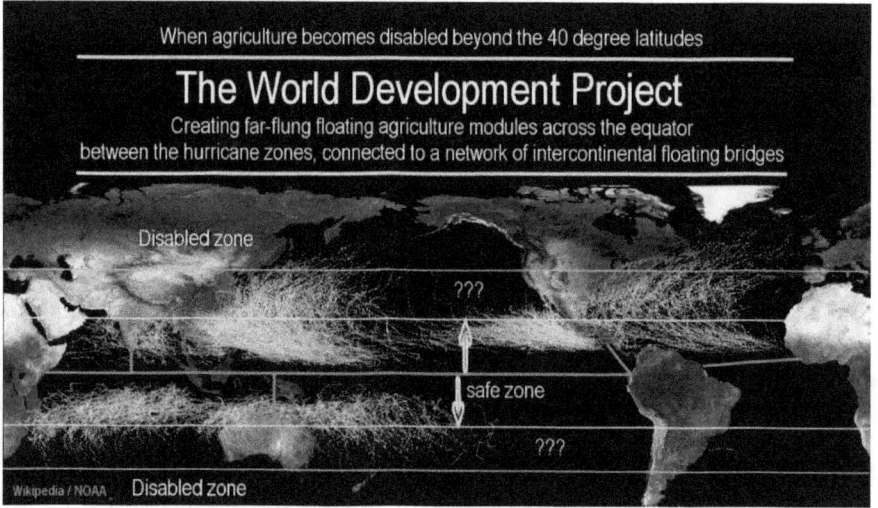

For as long as the recognition of the truth about the Sun remains blocked, the infrastructures will not be created that are needed to enable the relocation of all the northern nations into the tropics, together with their agriculture, when their territory becomes uninhabitable under the inactive Ice Age Sun. When the building of the needed infrastructures remains blocked, or long delayed, more than 90% of humanity is doomed to starve to death when the Ice Age transition occurs.

I expect that the truth about our Sun, which is actually rather simple and is already understood scientifically, will be acknowledged in time sufficiently to prevent the greatest potential universal suicide in human history. It is as simple as that. Our hope, therefore, literally rests in the fundamental nature of our humanity that includes a profound love for the truth.

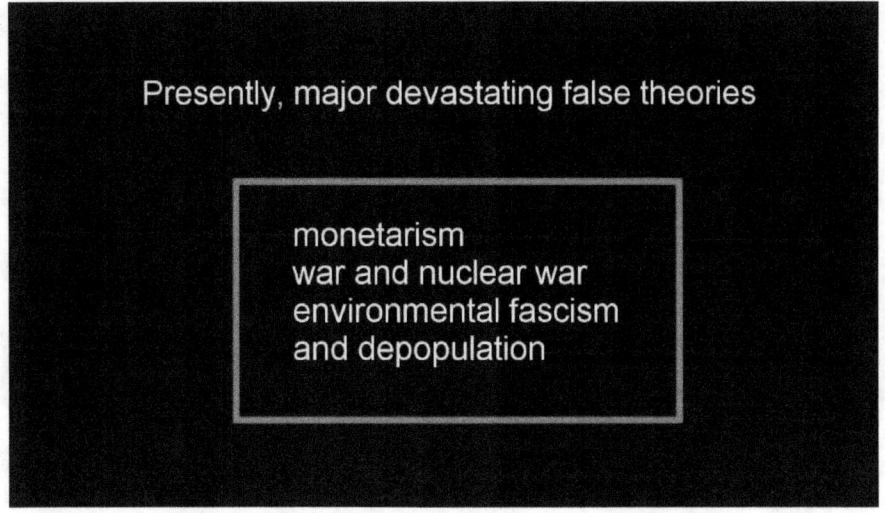

In fact, the recognition of the great Ice Age Challenge before us, which depends on a truthful recognition of the nature of our Sun, is of critical importance for the healing of society of its faith in the presently major devastating false theories in the political world that are destroying our precious humanity and civilization. These false theories are called, monetarism, war and nuclear war, environmental fascism, and depopulation. None of these theories stand on a single element of truth. Nor will these theories be overturned by technical adjustments in the form of laws. The games of adjustment of the theories, do not eradicate the lack of truth in these theories, but hide the lack thereof.

The great magnitude of the Ice Age Challenge

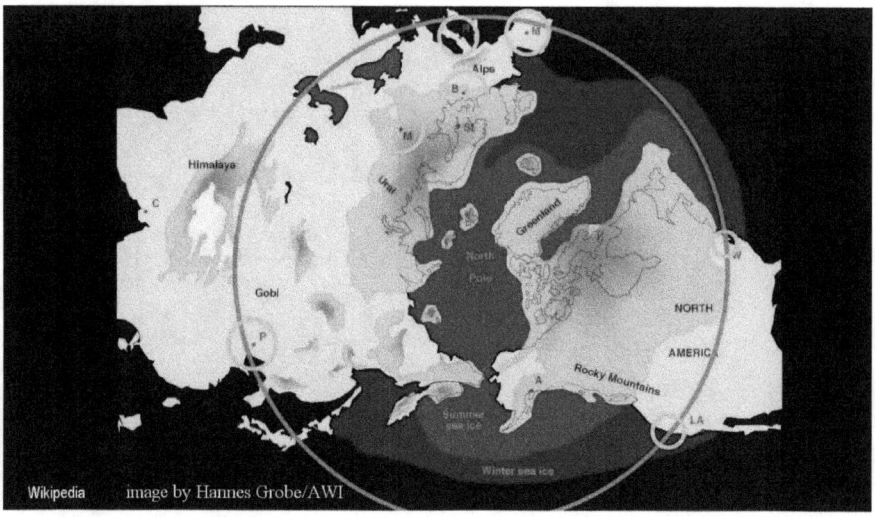

Wikipedia   image by Hannes Grobe/AWI

Hopefully, the great magnitude of the Ice Age Challenge and the focus on the truth that it inspires, will be sufficient at last to break the log-jam against the truth that all the other dangerous false theories are built on that are presently destroying civilization and threaten human extinction, as in the case of nuclear war.

With the power of the mind

Photo by Scott Williams

The Supreme Being

wikipedia

But, with the power of the mind, which is inherent in us as human beings, we have the means at hand to lift ourselves above the fog of faith in fairy tales and reorient the world onto the path of universal freedom in a renaissance of the truth. Why should we not be able to do this 'small' thing? Seeing the truth with the mind is native to our humanity. When a slave boy can understand absolute truth, as in Plato's story of the Meno dialog, then, obviously, so can we all.

www.ingramcontent.com/pod-product-compliance
Lightning Source LLC
Chambersburg PA
CBHW070228210526
45169CB00023B/1260